Copyright © 2023 by M.J. Knightly (Author)

This book is protected by copyright law and is intended solely for personal use. Reproduction, distribution, or any other form of use requires the written permission of the author. The information presented in this book is for educational and entertainment purposes only, and while every effort has been made to ensure its accuracy and completeness, no guarantees are made. The author is not providing legal, financial, medical, or professional advice, and readers should consult with a licensed professional before implementing any of the techniques discussed in this book. The content in this book has been sourced from various reliable sources, but readers should exercise their own judgment when using this information. The author is not responsible for any losses, direct or indirect, that may occur from the use of this book, including but not limited to errors, omissions, or inaccuracies.

We hope this book has been informative and helpful on your journey to understanding and celebrating older adults. Thank you for your interest and support!

Title: Evolution of Clean Energy: The Rise of Electric Vehicles
Subtitle: Battery Breakthroughs and Sustainable Mobility

Series: Ride Through Time: The Story of World Vehicles
By M.J. Knightly

"The transportation industry is on the brink of a revolution, with advancements in autonomous vehicles, electric cars, and new modes of transportation like hyperloops and vertical takeoff and landing aircrafts."
Mary Barra, CEO of General Motors

"Innovation in transportation is not only about speed and efficiency, but also about sustainability and reducing our impact on the environment."
Patricia Espinosa, Executive Secretary of the United Nations Framework Convention on Climate Change

"The future of transportation will be about creating a seamless and integrated experience, where different modes of transportation work together to get people where they need to go."
Dara Khosrowshahi, CEO of Uber

"The age of electric vehicles is here, and it's exciting to see how technology is driving the transformation of our transportation systems."
Elon Musk, CEO of Tesla and SpaceX

"Transportation is not just about getting from point A to point B, it's about the freedom and opportunities that come with mobility. Innovations in transportation are critical for empowering individuals and communities around the world."
Ban Ki-moon, former Secretary-General of the United Nations

"As transportation continues to evolve, we have the opportunity to create a more equitable and accessible system that benefits all people, regardless of their background or circumstances."
Keisha Lance Bottoms, former mayor of Atlanta, Georgia

"Innovation is key to unlocking the potential of transportation, and we need to encourage and invest in new ideas and technologies that will shape the future of mobility."
Hiroto Saikawa, former CEO of Nissan Motor Company

Table of Contents

Introduction
The state of the automotive industry

The automotive industry has been a driving force of technological innovation and economic growth for over a century. Today, however, the industry is facing a number of challenges that are fundamentally reshaping the way we think about transportation. One of the most significant of these challenges is the need to reduce greenhouse gas emissions and address climate change. In this context, electric vehicles (EVs) have emerged as a promising alternative to traditional gasoline-powered cars.

The state of the automotive industry:

The automotive industry is a complex and multifaceted sector that encompasses a wide range of companies, products, and services. At its core, however, the industry is defined by the production and sale of cars, trucks, and other vehicles. In recent years, the industry has experienced a number of significant trends and shifts that are transforming the way it operates.

One of the most important of these trends is the shift towards electric vehicles. In 2020, global EV sales reached a record high of over 3 million vehicles, representing a 43% increase from the previous year. While EVs still represent a relatively small share of the overall automotive market, their growth has been accelerating rapidly in recent years, and many experts predict that they will continue to gain market share in the coming years.

At the same time, the traditional internal combustion engine (ICE) remains the dominant technology in the automotive industry, accounting for over 90% of all vehicles sold worldwide. However, this is starting to change as governments around the world introduce policies to reduce greenhouse gas emissions and incentivize the adoption of cleaner technologies.

In addition to the shift towards electric vehicles, the automotive industry is also experiencing a number of other important trends and changes. These include the growth of autonomous driving technologies, the rise of ride-sharing and mobility-as-a-service platforms, and the increasing importance of sustainability and environmental responsibility.

Overall, the state of the automotive industry is complex and rapidly evolving. While traditional ICE vehicles continue to dominate the market, the rise of electric vehicles and other new technologies is fundamentally reshaping the industry and creating new opportunities and challenges for companies and consumers alike.

The rise of electric vehicles (EVs) is one of the most significant trends in the automotive industry today. As concerns over climate change and air pollution continue to mount, many consumers and policymakers are turning to EVs as a cleaner and more sustainable alternative to traditional gasoline-powered cars. In this section, we will explore the factors driving the rise of EVs and the implications of this trend for the future of transportation.

The rise of electric vehicles:

The rise of electric vehicles can be traced back to a number of factors, including advances in battery technology, government incentives and regulations, and changing consumer attitudes towards sustainability and environmental responsibility.

One of the key drivers of the rise of EVs is the rapid improvement in battery technology. In recent years, the cost of lithium-ion batteries, which are the most common type of battery used in EVs, has fallen dramatically, making EVs more affordable for consumers. At the same time, the range and performance of EVs have improved significantly, with many models now able to travel over 200 miles on a single charge.

Another important factor driving the rise of EVs is government incentives and regulations. Many countries and regions around the world have introduced policies to encourage the adoption of EVs, such as tax credits, rebates, and subsidies. In addition, some governments have introduced regulations that require automakers to produce a certain percentage of zero-emission vehicles, which has helped to spur innovation and investment in EV technology.

Changing consumer attitudes towards sustainability and environmental responsibility have also played a role in the rise of EVs. As more people become aware of the negative impacts of climate change and air pollution, they are increasingly looking for ways to reduce their carbon footprint and make more sustainable choices. EVs, with their zero-emissions technology, offer a compelling solution to these concerns.

Implications for the future of transportation:

The rise of electric vehicles has significant implications for the future of transportation. As more consumers adopt EVs, we can expect to see a shift away from traditional gasoline-powered cars, which could have a major impact on the global economy and the environment.

One of the most significant implications of the rise of EVs is the potential to reduce greenhouse gas emissions and mitigate the impacts of climate change. If EVs were to replace all traditional ICE vehicles on the road, it could reduce global CO2 emissions by up to 1.5 billion tons per year. This would represent a significant step towards achieving the emissions reductions targets outlined in the Paris Agreement on climate change.

In addition, the rise of EVs has the potential to transform the energy sector, by increasing the demand for renewable energy sources such as solar and wind power. As more EVs are introduced onto the grid, there will be a need for more charging infrastructure, which could drive investment in renewable energy and energy storage technologies.

Finally, the rise of EVs has important implications for the automotive industry itself. As EVs become more mainstream, we can expect to see a shift in the competitive landscape, with new players entering the market and traditional automakers adapting to the new reality. This could lead to new business models, such as subscription-based EV services, and increased innovation in areas such as autonomous driving and vehicle-to-grid technology.

Conclusion:

The rise of electric vehicles is a transformative trend that is reshaping the automotive industry and the way we think about transportation. By driving innovation and investment in clean energy technologies, EVs have the potential to help mitigate the impacts of climate change and create a more sustainable future for all. However, achieving this vision will require continued investment and collaboration from policymakers, businesses, and consumers around the world.

The importance of sustainable transportation

Sustainable transportation is a critical element in mitigating the adverse impacts of climate change, reducing pollution, and enhancing the quality of life for people. The transportation sector is responsible for a significant portion of global greenhouse gas emissions, primarily from the use of fossil fuels in vehicles. As such, transitioning to sustainable transportation is a crucial step in achieving global climate goals.

Electric vehicles represent a significant opportunity to shift away from fossil fuel-powered transportation and towards a more sustainable future. Unlike traditional gasoline-powered vehicles, electric vehicles produce no direct emissions and can be powered by renewable energy sources like wind or solar power. Additionally, electric vehicles are significantly more energy-efficient than traditional combustion engines, which means they require less energy to operate and can travel further on the same amount of energy.

By promoting sustainable transportation, we can create a cleaner, healthier, and more sustainable future for ourselves and future generations. Sustainable transportation is not just about electric vehicles, but it also includes other modes of transportation like public transit, biking, and walking. By creating more walkable and bikeable cities, we can reduce reliance on cars and promote healthier lifestyles.

Sustainable transportation also has broader societal benefits beyond just reducing emissions. By improving access to transportation options, we can enhance mobility and reduce social inequalities. Sustainable transportation can also support local economies by creating jobs in industries such as electric vehicle manufacturing, renewable energy, and public transit.

Overall, the importance of sustainable transportation cannot be overstated. As we move towards a more sustainable future, we must prioritize sustainable transportation as a crucial component of our efforts to mitigate climate change and create a more equitable, healthy, and prosperous society.

Chapter 1: The History of Electric Vehicles

The early history of electric vehicles

Electric vehicles have been around for over a century, and their early history is both fascinating and complex. While many people assume that electric vehicles are a recent innovation, the truth is that they predate gasoline-powered vehicles by several decades.

The first electric vehicle was built in 1832 by Scottish inventor Robert Anderson. This early electric carriage was powered by non-rechargeable primary cells and was used primarily for short trips around Edinburgh. However, due to the limitations of battery technology at the time, it was not a practical mode of transportation for most people.

It wasn't until the late 1800s that electric vehicles began to gain traction as a viable transportation option. In 1891, William Morrison, an American inventor, developed the first electric vehicle capable of traveling significant distances. His six-passenger vehicle could travel up to 14 miles on a single charge and could reach a top speed of around 14 miles per hour. This breakthrough in battery technology paved the way for the development of more practical electric vehicles.

By the turn of the century, electric vehicles had become a popular mode of transportation in urban areas.

Electric taxis, buses, and delivery vehicles could be found in cities across Europe and North America. In 1900, electric vehicles made up around one-third of all vehicles on the road in the United States, with gasoline and steam-powered vehicles making up the remainder.

Electric vehicles offered several advantages over their gasoline-powered counterparts. They were quiet, emitted no exhaust fumes, and were easy to operate. Additionally, electric vehicles required less maintenance than gasoline-powered vehicles, which made them an attractive option for urban delivery fleets.

However, the early success of electric vehicles was short-lived. The invention of the electric starter motor in 1912 made gasoline-powered vehicles more practical for everyday use. Additionally, advancements in gasoline engine technology made gasoline-powered vehicles more efficient and reliable, which made them more attractive to consumers.

By the 1920s, electric vehicles had fallen out of favor, and production of electric vehicles had all but ceased. Despite this setback, electric vehicles continued to be used in niche applications, such as forklifts, golf carts, and some industrial vehicles.

It wasn't until the 1960s that electric vehicles began to make a comeback. The rising concerns over air pollution and energy security led to renewed interest in electric vehicles as a cleaner, more sustainable mode of transportation. In 1966, General Motors developed the first modern electric car, the Electrovair II, which was powered by a 115-horsepower electric motor and a lead-acid battery pack. While the Electrovair II was never produced for mass consumption, it paved the way for the development of more advanced electric vehicles in the years to come.

In the following decades, electric vehicles continued to make slow but steady progress. The development of more advanced battery technologies, such as nickel-metal hydride and lithium-ion batteries, made electric vehicles more practical and efficient. In the 1990s, the first mass-produced electric vehicle, the GM EV1, was introduced in California. While the EV1 was ultimately discontinued, it paved the way for the development of more practical and affordable electric vehicles in the years to come.

Today, electric vehicles are poised to transform the transportation industry. With advancements in battery technology, electric vehicles are becoming more affordable, more efficient, and more practical for everyday use. As we look towards the future, it's clear that electric vehicles will play a critical role in reducing greenhouse gas emissions, improving air quality, and promoting a more sustainable transportation system.

The decline of electric vehicles

During the early 20th century, electric vehicles had become quite popular in the United States and Europe due to their low noise levels, ease of operation, and lack of emissions. In fact, by 1912, electric vehicles made up around 38% of all cars on US roads. However, their dominance was short-lived, and by the 1920s, gasoline-powered vehicles had become the primary mode of transportation. There were several reasons for this decline, including:

1. Limited driving range: One of the main disadvantages of early electric vehicles was their limited driving range, typically around 50 miles per charge. This made them impractical for long-distance travel or even for daily use by some consumers.

2. High cost: Electric vehicles were expensive to produce, and as a result, they were priced out of reach for most consumers. In contrast, gasoline-powered vehicles were mass-produced and became increasingly affordable.

3. Lack of infrastructure: Unlike gasoline-powered vehicles, which could be refueled at gas stations, electric vehicles required charging stations, which were not widely available. As a result, electric vehicle owners had to charge their vehicles at home, which limited their range even further.

4. Emergence of the gasoline-powered vehicle: As gasoline-powered vehicles became more widely available, they offered several advantages over electric vehicles, including longer driving range, lower cost, and a more established infrastructure.

5. Improvements in road networks: With the rise of paved roads and highways, gasoline-powered vehicles became more practical for long-distance travel, while electric vehicles struggled to keep up.

As a result of these factors, electric vehicles fell out of favor and were largely forgotten for several decades. However, with the growing concern over air pollution, greenhouse gas emissions, and climate change, electric vehicles have once again become a focus of research and development. Today, electric vehicles are viewed as a key component of a sustainable transportation system, and efforts are being made to address many of the challenges that led to their decline in the past.

The resurgence of electric vehicles in the modern era

The modern era has seen a resurgence in electric vehicles, driven by concerns over climate change, air pollution, and energy security. The following are some key factors that have contributed to the renewed interest in electric vehicles:

1. Advances in battery technology: One of the biggest hurdles for electric vehicles has been the limitations of battery technology. However, in recent years, significant advancements have been made in this area, resulting in batteries that are lighter, more efficient, and have longer driving ranges. This has made electric vehicles more practical for everyday use and has helped to overcome one of the major barriers to their adoption.

2. Government support: Many governments around the world have implemented policies and incentives to promote the adoption of electric vehicles. These include tax credits, rebates, and grants for consumers and businesses that purchase electric vehicles, as well as funding for research and development in this area. This support has helped to create a market for electric vehicles and has incentivized manufacturers to invest in this technology.

3. Environmental concerns: With growing concerns over air pollution and climate change, there has been a renewed focus on finding alternative, cleaner forms of transportation. Electric vehicles produce no emissions at the tailpipe, and when charged using renewable energy sources, they can significantly reduce greenhouse gas emissions. This has led to increased interest in electric vehicles as a more sustainable form of transportation.

4. Consumer demand: As awareness of electric vehicles has grown, so too has consumer interest in these vehicles. Many consumers are drawn to the lower operating costs of electric vehicles, as well as their quiet operation and smooth ride. In addition, the growing availability of charging infrastructure has made electric vehicles more practical for everyday use, further increasing demand.

5. Technological advancements: In addition to advances in battery technology, there have been many other technological advancements that have helped to make electric vehicles more practical and efficient. These include improvements in motors, power electronics, and energy management systems, as well as the development of regenerative braking and other energy-saving technologies.

The resurgence of electric vehicles in the modern era has been a significant development in the automotive industry, and one that has the potential to transform transportation in the years to come. As manufacturers continue to invest in this technology and governments provide support, we can expect to see continued growth in the adoption of electric vehicles and a shift towards a more sustainable transportation system.

Chapter 2: Types of Electric Vehicles
Battery electric vehicles (BEVs)

Battery electric vehicles (BEVs) are one of the most popular types of electric vehicles (EVs) available today. Unlike hybrid vehicles, which use both an internal combustion engine and an electric motor, BEVs rely solely on electric power. In this chapter, we'll take a closer look at BEVs and explore how they work, their benefits and drawbacks, and their potential to transform transportation in the future.

How do BEVs work? BEVs are powered by electric motors that are powered by rechargeable batteries. These batteries are typically made up of a series of lithium-ion cells, similar to those found in laptops and smartphones. The batteries are charged by plugging the vehicle into an electrical outlet or charging station. BEVs have a range that varies depending on the battery size and vehicle model, but most modern BEVs can travel around 200-300 miles on a single charge.

Benefits of BEVs One of the primary benefits of BEVs is their reduced environmental impact compared to traditional gasoline-powered vehicles. BEVs produce zero emissions at the tailpipe, which can help reduce air pollution and greenhouse gas emissions. In addition, BEVs are typically quieter and smoother to drive than gasoline-powered vehicles, and they require less maintenance due to their simpler drivetrains.

BEVs can also be more cost-effective than traditional gasoline-powered vehicles over the long term. While the initial cost of a BEV can be higher than a traditional vehicle, the cost of fuel and maintenance is typically lower. Electricity is generally less expensive than gasoline, and BEVs require less frequent maintenance since they have fewer moving parts.

Finally, BEVs can also provide a more enjoyable driving experience. Electric motors offer instant torque, which can make acceleration feel more responsive and exciting. In addition, many BEVs offer advanced features such as regenerative braking, which can help extend the vehicle's range and provide a more efficient driving experience.

Drawbacks of BEVs Despite their many benefits, BEVs also have some drawbacks that are important to consider. One of the biggest concerns for potential BEV owners is range anxiety - the fear that the vehicle will run out of power before reaching its destination. While the range of BEVs is improving, it can still be limiting for some drivers, particularly those who frequently travel long distances.

In addition, the charging infrastructure for BEVs is still developing, and many areas have limited public charging stations. This can make it difficult for BEV owners to find a place to charge their vehicle while on the go. Charging times for BEVs can also be longer than refueling times for gasoline-powered vehicles, which can be an inconvenience for some drivers.

Finally, BEVs are not suitable for all driving situations. They may not be practical for drivers who frequently tow heavy loads or travel off-road. In addition, extreme temperatures can impact the performance of BEV batteries, which may limit their use in certain climates.

The Future of BEVs Despite their drawbacks, BEVs are becoming increasingly popular as more drivers look for environmentally friendly and cost-effective transportation options. Many automakers are investing heavily in the development of BEVs, and new models are being released every year. As battery technology continues to improve, the range of BEVs is likely to increase, which could help address concerns about range anxiety. In addition, the charging infrastructure for BEVs is also expanding, which should make it easier for BEV owners to find a place to charge their vehicle while on the go.

In the future, BEVs could play a key role in reducing air pollution and greenhouse gas emissions from transportation. As more renewable energy sources such as wind and solar power come online, BEVs could be charged using cleaner energy, further reducing their environmental impact.

Hybrid electric vehicles (HEVs) are a type of electric vehicle that combines an internal combustion engine (ICE) with an electric motor and battery. The primary purpose of this combination is to increase fuel efficiency and reduce emissions.

HEVs have been around since the late 1990s and have been growing in popularity due to their improved efficiency and reduced emissions. They work by using the electric motor to supplement the ICE, allowing the vehicle to use less fuel and emit fewer pollutants. The electric motor also acts as a generator, recovering energy during braking and deceleration and storing it in the battery for later use.

There are two main types of HEVs: parallel and series. In a parallel hybrid, both the ICE and electric motor are connected to the transmission and can drive the wheels directly. In a series hybrid, the ICE is used solely to generate electricity to power the electric motor, which then drives the wheels.

HEVs can also be further classified as mild or full hybrids. Mild hybrids use a smaller electric motor and battery and cannot drive solely on electric power. Full hybrids, on the other hand, have larger electric motors and batteries and can operate in electric-only mode for short distances at low speeds.

One example of a popular HEV is the Toyota Prius, which was introduced in Japan in 1997 and later launched globally in 2000. The Prius uses a series-parallel hybrid system that allows it to operate in electric-only mode at low speeds and during deceleration. It also features regenerative braking, which allows it to recover energy during braking and store it in the battery for later use.

Another example of an HEV is the Honda Insight, which was first introduced in the United States in 1999. The Insight uses a parallel hybrid system that allows both the ICE and electric motor to drive the wheels directly. It also features a lightweight aluminum body and a low coefficient of drag, which further improve its fuel efficiency.

Overall, HEVs are an important type of electric vehicle that offer improved fuel efficiency and reduced emissions compared to traditional gasoline-powered vehicles. As battery technology continues to improve, it is possible that HEVs may eventually be replaced by more advanced types of electric vehicles, such as plug-in hybrid electric vehicles (PHEVs) and battery electric vehicles (BEVs). However, for now, they remain a popular and effective way to reduce the environmental impact of transportation.

Plug-in hybrid electric vehicles (PHEVs) are another type of electric vehicle that has gained popularity in recent years. PHEVs combine an internal combustion engine with an electric motor and battery pack. This combination allows the vehicle to operate in electric mode, gasoline mode, or a combination of both.

One of the primary advantages of PHEVs is their extended range. Unlike BEVs, which rely solely on their battery packs, PHEVs have a gasoline engine that can provide additional range when the battery is depleted. This means that PHEVs can be driven for longer distances than BEVs without needing to stop and recharge.

Another advantage of PHEVs is their flexibility. They can be charged using a standard household outlet, a dedicated charging station, or they can be refueled with gasoline. This makes PHEVs a good choice for drivers who may have concerns about range anxiety, or who want the flexibility to switch between gasoline and electric power depending on their driving needs.

There are several different types of PHEVs available on the market, with different battery sizes and ranges. Some PHEVs have smaller battery packs and shorter electric ranges, while others have larger battery packs and can travel further on electric power alone.

One of the challenges of PHEVs is their cost. They are generally more expensive than comparable gasoline-only vehicles, although government incentives and tax credits can help offset some of the cost. In addition, PHEVs require more complex powertrain systems than traditional gasoline vehicles, which can lead to higher maintenance costs over the life of the vehicle.

Despite these challenges, PHEVs have become an increasingly popular choice for drivers looking for a vehicle that offers the best of both worlds: the efficiency and low emissions of electric power, and the extended range and flexibility of gasoline power. As battery technology continues to improve, PHEVs are likely to become even more efficient and affordable, making them a compelling option for drivers around the world.

Fuel cell electric vehicles (FCEVs)

Fuel cell electric vehicles (FCEVs) are a type of electric vehicle that uses hydrogen fuel cells to generate electricity to power an electric motor. FCEVs have the potential to be an important part of the transition to sustainable transportation, as they produce no harmful emissions and have longer ranges than battery electric vehicles (BEVs).

History of FCEVs:

The history of fuel cell technology dates back to the early 19th century, when the first fuel cell was developed by Sir William Grove. However, it was not until the 1960s that fuel cell technology began to be developed for use in transportation.

The first fuel cell vehicle was developed in 1959 by General Electric, and the first fuel cell-powered car was developed by Ford in 1966. In the following decades, fuel cell technology continued to be developed, but progress was slow due to technical challenges and high costs.

In the 2000s, several automakers began to develop fuel cell vehicles for commercial use. The first commercially available fuel cell vehicle was the Toyota Mirai, which was released in Japan in 2014.

How FCEVs work:

FCEVs use a fuel cell to generate electricity from hydrogen and oxygen. The fuel cell produces electricity through an electrochemical reaction, which is used to power an electric motor.

The hydrogen fuel is stored in high-pressure tanks, similar to the way gasoline is stored in a traditional car. When the car is in use, the hydrogen fuel is delivered to the fuel cell, where it reacts with oxygen from the air to produce electricity and water vapor.

Advantages of FCEVs:

1. Zero emissions: FCEVs produce no harmful emissions, making them a clean transportation option.

2. Long range: FCEVs have longer ranges than BEVs, typically around 300-400 miles per tank of hydrogen.

3. Quick refueling: FCEVs can be refueled in a matter of minutes, similar to the time it takes to refuel a gasoline-powered car.

4. High efficiency: FCEVs are highly efficient, with a fuel economy of around 60 miles per kilogram of hydrogen.

Challenges facing FCEVs:

1. Cost: FCEVs are currently more expensive than BEVs or traditional gasoline-powered cars due to the high cost of the fuel cell technology and the infrastructure needed to produce and distribute hydrogen.

2. Limited infrastructure: The infrastructure for producing and distributing hydrogen fuel is currently limited, which can make it difficult to refuel FCEVs.

3. Safety concerns: Hydrogen is highly flammable and requires special safety measures for handling and storage.

Conclusion:

Fuel cell electric vehicles have the potential to play an important role in the transition to sustainable transportation. While they currently face challenges related to cost and infrastructure, ongoing research and development could help to overcome these obstacles and make FCEVs a viable alternative to traditional gasoline-powered cars.

Chapter 3: Electric Vehicle Components
Battery technology and charging infrastructure

Battery technology and charging infrastructure are two key components that make electric vehicles function. This chapter will explore the advancements in battery technology and the development of charging infrastructure to support the growth of the electric vehicle industry.

Battery Technology The battery is the heart of an electric vehicle, providing the energy needed to power the motor. Advances in battery technology have been critical to the success of electric vehicles, as they have allowed for greater driving ranges and reduced charging times.

One of the most significant advancements in battery technology has been the development of lithium-ion batteries, which have a higher energy density than previous battery types. This means that they can store more energy in a smaller space, allowing for lighter and more efficient electric vehicles. In addition, lithium-ion batteries have longer lifetimes and can be charged more quickly than other battery types.

However, lithium-ion batteries are not without their limitations. They are still relatively expensive to manufacture and can pose safety risks if not handled properly. As a result, researchers are exploring new battery chemistries and materials that could provide even greater energy density, longer lifetimes, and improved safety.

Charging Infrastructure Charging infrastructure is essential for electric vehicle owners, as it provides the means to recharge their vehicle's battery. The development of a robust charging infrastructure is crucial for the widespread adoption of electric vehicles, as it ensures that drivers have access to charging stations wherever they go.

There are three main types of charging infrastructure: Level 1, Level 2, and DC fast charging. Level 1 charging is the slowest method, as it uses a standard household outlet and takes several hours to fully charge an electric vehicle. Level 2 charging uses a higher voltage and can charge an electric vehicle in a few hours. DC fast charging is the fastest method and can charge an electric vehicle to 80% capacity in as little as 30 minutes.

Governments, automakers, and other stakeholders are investing in the development of charging infrastructure to support the growth of the electric vehicle industry. In many countries, governments are providing funding to build public charging stations, while automakers are partnering with charging companies to develop their own networks. In addition, new technologies are being developed to make charging more convenient and accessible, such as wireless charging pads and charging robots.

Conclusion Battery technology and charging infrastructure are two critical components of electric vehicles. Advances in battery technology have made electric vehicles more efficient and practical, while the development of a robust charging infrastructure is essential for the widespread adoption of electric vehicles. As the electric vehicle industry continues to grow, it is likely that we will see further advancements in battery technology and charging infrastructure to support this growth.

Electric vehicles (EVs) rely on several key components to power their propulsion system, including the motor and the power electronics that control it. The motor converts electrical energy from the battery into mechanical energy, which drives the wheels and propels the vehicle forward. The power electronics, meanwhile, regulate the flow of electricity between the battery and the motor, ensuring that the vehicle operates efficiently and safely. In this chapter, we'll take a closer look at these critical components and explore their role in the electric vehicle powertrain.

Electric Motors

The electric motor is the heart of the electric vehicle powertrain, providing the motive force to drive the wheels. Unlike internal combustion engines, which rely on combustion of fuel to generate power, electric motors use magnetic fields to convert electrical energy into rotational energy. This makes them highly efficient, producing more torque at lower speeds and consuming less energy overall than gasoline or diesel engines.

There are several different types of electric motors used in electric vehicles, each with its own unique characteristics and performance capabilities. Some of the most common types include:

1. Permanent Magnet Motors: These motors use permanent magnets to create a magnetic field that interacts with a rotating magnetic field produced by the motor's stator. This interaction produces torque, which drives the wheels. Permanent magnet motors are simple, efficient, and reliable, making them a popular choice for many electric vehicle manufacturers.

2. Induction Motors: Also known as asynchronous motors, these motors use electromagnetic induction to create a magnetic field that interacts with the stator's rotating magnetic field. Induction motors are reliable and easy to manufacture, but they tend to be less efficient than permanent magnet motors at low speeds.

3. Switched Reluctance Motors: These motors use electromagnets to generate a magnetic field that interacts with the stator's magnetic field. Switched reluctance motors are highly efficient and reliable, but they can be noisy and produce significant amounts of vibration.

4. Brushless DC Motors: These motors use electronic controls to create a rotating magnetic field that interacts with a stationary magnet. Brushless DC motors are highly efficient and reliable, with no brushes to wear out or require maintenance. They're also relatively quiet and produce little vibration.

Power Electronics

In addition to the motor, electric vehicles also rely on a complex system of power electronics to regulate the flow of electricity between the battery, the motor, and other components. This system includes several key components, including:

1. Inverter: The inverter is responsible for converting the DC power stored in the battery into the AC power needed to drive the motor. It also regulates the flow of power to the motor to ensure that it operates safely and efficiently.

2. DC/DC Converter: The DC/DC converter is responsible for converting the high voltage DC power from the battery to the lower voltage DC power needed to operate other vehicle systems, such as the lights and air conditioning.

3. Charger: The charger is responsible for converting AC power from the electrical grid to DC power that can be stored in the vehicle's battery. It also regulates the charging process to ensure that the battery is charged safely and efficiently.

4. Battery Management System: The battery management system (BMS) is responsible for monitoring the state of the battery, including its temperature, voltage, and current. It also regulates the charging and discharging

process to ensure that the battery operates safely and efficiently.

Conclusion

The motor and power electronics are critical components of the electric vehicle powertrain, working together to provide efficient, reliable, and low-emission propulsion. As electric vehicle technology continues to evolve, we can expect to see further advancements in motor and power electronics technology, leading to more efficient and affordable electric vehicles for consumers.

Regenerative braking

Regenerative braking is an innovative technology used in electric and hybrid vehicles that allows the vehicle to recover some of the energy lost during braking. This technology can significantly increase the vehicle's efficiency and extend its driving range.

When a conventional vehicle brakes, the kinetic energy of the moving vehicle is converted into heat energy, which is dissipated through the brake pads and discs. In contrast, regenerative braking uses the electric motor of the vehicle to slow down the vehicle and convert its kinetic energy into electrical energy, which is then stored in the battery for later use.

Regenerative braking works by using the electric motor of the vehicle as a generator. When the driver applies the brakes, the electric motor is engaged in reverse, and it converts the kinetic energy of the vehicle into electrical energy. The electrical energy is then sent back to the battery, where it is stored for later use.

There are several benefits to regenerative braking. First, it can significantly increase the efficiency of electric and hybrid vehicles. By recovering some of the energy lost during braking, regenerative braking can increase the vehicle's driving range and reduce its energy consumption.

Second, regenerative braking can improve the overall driving experience of electric and hybrid vehicles. Since regenerative braking allows the vehicle to slow down smoothly and efficiently, it can provide a more comfortable and seamless driving experience.

Finally, regenerative braking can help to reduce the wear and tear on the brakes of electric and hybrid vehicles. Since the electric motor is used to slow down the vehicle, the traditional braking system is used less frequently, which can extend the lifespan of the brake pads and discs.

Despite its many benefits, regenerative braking is not without its limitations. For example, regenerative braking is less effective at low speeds and in stop-and-go traffic, where the vehicle is constantly accelerating and decelerating. In addition, regenerative braking can only recover a portion of the energy lost during braking, so it is important for electric and hybrid vehicle drivers to practice efficient driving habits to maximize their vehicle's efficiency.

Overall, regenerative braking is an important technology in the world of electric and hybrid vehicles. By recovering some of the energy lost during braking, regenerative braking can help to increase the efficiency of electric and hybrid vehicles, provide a more comfortable driving experience, and reduce the wear and tear on the brakes. As electric and hybrid vehicles continue to gain popularity, regenerative braking will play an increasingly important role in the future of transportation.

Electric vehicles (EVs) rely on a sophisticated energy management system (EMS) to ensure that they operate efficiently, reliably, and safely. The EMS is responsible for monitoring and controlling the flow of energy between the battery, motor, and other components of the vehicle. This includes managing the charging and discharging of the battery, controlling the power output of the motor, and regulating the flow of electricity throughout the vehicle's electrical system.

The key components of an EMS include the battery management system (BMS), the power electronics module (PEM), and the vehicle control unit (VCU). The BMS is responsible for monitoring the state of the battery, including its voltage, temperature, and state of charge. It also controls the charging and discharging of the battery, ensuring that it is charged to the appropriate level and that it is not discharged beyond its safe limit. The PEM is responsible for controlling the power output of the motor, converting the DC power from the battery into AC power that can be used by the motor. The VCU is responsible for controlling the overall operation of the vehicle, including the energy management system, motor, and other components.

One of the most important functions of an EMS is to maximize the efficiency of the vehicle. This is achieved through a process known as regenerative braking, which involves using the motor as a generator to capture energy that would otherwise be lost during braking. When the driver applies the brakes, the motor is used to slow the vehicle down, and the energy generated during this process is stored in the battery for later use. This not only helps to extend the range of the vehicle but also reduces the wear and tear on the brakes, leading to lower maintenance costs.

Another important function of an EMS is to manage the charging of the battery. This includes controlling the rate of charge, monitoring the state of charge, and ensuring that the battery is not overcharged or discharged beyond its safe limit. The EMS must also take into account factors such as the temperature and state of the battery, as these can affect its performance and lifespan.

In addition to managing the battery and motor, an EMS must also control the flow of electricity throughout the vehicle's electrical system. This includes managing the power usage of other components such as the air conditioning, lighting, and infotainment system. The EMS must also ensure that the electrical system is properly grounded and protected against short circuits and other electrical faults.

Overall, the energy management system is a critical component of an electric vehicle, ensuring that it operates efficiently, reliably, and safely. As EV technology continues to evolve, we can expect to see further advances in energy management systems, including improved battery management, more efficient power electronics, and more sophisticated control algorithms. These advances will help to extend the range of electric vehicles, reduce their cost, and make them an even more attractive option for drivers around the world.

Chapter 4: The Environmental Impact of Electric Vehicles

The benefits of reducing emissions

As the world becomes increasingly aware of the negative impact of greenhouse gas emissions, there has been a growing interest in reducing the amount of pollutants that are released into the atmosphere. One of the most significant contributors to this problem is transportation, particularly the burning of fossil fuels in vehicles. Electric vehicles (EVs) have emerged as a potential solution to this issue, offering a cleaner and more sustainable way to travel. In this section, we will explore the benefits of reducing emissions through the use of EVs.

1. Reduction in Greenhouse Gas Emissions The most significant benefit of EVs is their ability to reduce greenhouse gas emissions. According to a study by the Union of Concerned Scientists, an average EV in the United States produces less than half the greenhouse gas emissions of a gasoline-powered vehicle. The reduction is even greater when the electricity used to charge the EV comes from renewable sources such as wind or solar power. By using electricity instead of gasoline, EVs can significantly reduce the amount of carbon dioxide, nitrogen oxides, and other pollutants that are released into the atmosphere.

2. Improved Air Quality The reduction in greenhouse gas emissions also translates into improved air quality. Burning fossil fuels in vehicles produces a wide range of pollutants, including particulate matter, carbon monoxide, and sulfur dioxide. These pollutants can have a severe impact on human health, particularly on those with pre-existing respiratory conditions. By replacing gasoline-powered vehicles with EVs, the amount of these pollutants in the air can be significantly reduced, leading to better overall air quality and a reduction in related health problems.

3. Lower Noise Pollution In addition to producing fewer emissions, EVs also create less noise pollution than traditional gasoline-powered vehicles. The electric motors in EVs operate much more quietly than internal combustion engines, reducing noise pollution in urban areas. This is particularly important in densely populated cities, where the constant noise of traffic can have a significant impact on the quality of life for residents.

4. Increased Energy Efficiency Another benefit of EVs is their increased energy efficiency compared to gasoline-powered vehicles. EVs use energy more efficiently, with up to 80% of the energy stored in their batteries converted into usable power, compared to only around 20% for gasoline engines. This increased efficiency means that EVs can travel

further on a single charge, reducing the need for frequent refueling stops and reducing overall energy consumption.

5. Lower Operating Costs Finally, EVs can offer lower operating costs than traditional gasoline-powered vehicles. The cost of electricity is typically much lower than gasoline, and maintenance costs are often lower as well, due to the simpler design of electric motors and the reduced wear and tear on brakes and other components. This can make EVs a more economical choice for many drivers, particularly for those who frequently drive in urban areas with heavy traffic.

In conclusion, there are many benefits to reducing emissions through the use of EVs. By producing fewer greenhouse gas emissions, improving air quality, reducing noise pollution, increasing energy efficiency, and lowering operating costs, EVs offer a sustainable and practical alternative to traditional gasoline-powered vehicles. As the world continues to face the challenges of climate change and environmental degradation, the transition to EVs represents a critical step towards a cleaner, more sustainable future.

Life cycle analysis of electric vehicles

The increasing global concern for climate change and environmental degradation has led to a growing interest in electric vehicles (EVs) as a potential solution to reduce the greenhouse gas emissions associated with transportation. While the use of EVs is generally recognized as a cleaner alternative to conventional gasoline-powered vehicles, it is important to understand the full life cycle environmental impact of EVs to evaluate their sustainability. Life cycle analysis (LCA) is a widely accepted method to assess the environmental impacts of products or systems throughout their life cycle, from raw material extraction to end-of-life disposal. In this chapter, we will examine the results of LCAs conducted on EVs to gain a better understanding of their environmental impact.

Life Cycle Analysis of Electric Vehicles:

An LCA of an EV includes the environmental impact associated with the production, use, and disposal of the vehicle and its components. It takes into account factors such as energy consumption, greenhouse gas emissions, air and water pollution, and resource depletion. The life cycle of an EV includes three main stages: the manufacturing stage, the use stage, and the end-of-life stage.

Manufacturing Stage:

The manufacturing stage includes the extraction and processing of raw materials, component manufacturing, and assembly. The main components of an EV include the battery pack, electric motor, power electronics, and other supporting systems. The battery pack is the most significant component of an EV in terms of its environmental impact. The production of batteries requires the extraction and processing of raw materials such as lithium, nickel, and cobalt, which can result in environmental impacts such as land degradation, water pollution, and carbon emissions. However, the environmental impact of battery production can be reduced by using recycled materials and increasing the efficiency of production processes.

The Use Stage:

The use stage of an EV includes the energy consumption and emissions associated with charging and operating the vehicle. EVs have lower emissions than conventional gasoline-powered vehicles during the use stage because they do not have a tailpipe emissions. However, the environmental impact of the use stage depends on the source of electricity used to charge the vehicle. In regions with a high percentage of renewable energy sources, such as wind and solar, EVs have significantly lower greenhouse gas emissions compared to conventional vehicles. However, in regions where the electricity mix is dominated by fossil fuels, the environmental benefits of EVs are reduced.

End-of-Life Stage:

The end-of-life stage of an EV includes the disposal or recycling of the vehicle and its components. EVs have a longer lifespan compared to conventional gasoline-powered vehicles, which can reduce the environmental impact of end-of-life disposal. The disposal of batteries is a significant environmental concern because batteries can contain hazardous materials such as lead and lithium. However, the recycling of batteries can reduce the environmental impact of disposal by recovering valuable materials and reducing the demand for new raw materials.

Conclusion:

LCA studies have shown that the environmental impact of EVs is significantly lower than conventional gasoline-powered vehicles, particularly during the use stage. The production of batteries and the source of electricity used to charge the vehicle can have a significant impact on the environmental sustainability of EVs. However, the environmental impact of EVs can be reduced by using recycled materials in battery production and increasing the share of renewable energy in the electricity mix. The recycling of batteries can also reduce the environmental impact of end-of-life disposal. Therefore, it is important to continue to develop sustainable production and end-of-life disposal practices to ensure the long-term environmental sustainability of EVs.

Comparing electric vehicles to other modes of transportation

Electric vehicles are just one of many modes of transportation available to us, including gasoline-powered vehicles, public transportation, walking, and biking. When considering the environmental impact of electric vehicles, it is important to compare them to these other modes of transportation to get a full picture of their benefits and limitations.

Gasoline-powered vehicles

Gasoline-powered vehicles have been the primary mode of transportation for decades, but they come with significant environmental impacts. Burning gasoline produces harmful emissions, including carbon dioxide, nitrogen oxides, and particulate matter, which contribute to air pollution and climate change. In addition, the extraction, refinement, and transportation of gasoline have their own environmental impacts, including habitat destruction, water pollution, and oil spills.

Compared to gasoline-powered vehicles, electric vehicles emit far fewer pollutants, which makes them a cleaner alternative. However, electric vehicles are only as clean as the energy they are powered by. If the electricity used to charge an electric vehicle comes from a coal-fired power plant, for example, the environmental benefits of the electric vehicle are greatly reduced.

Public transportation

Public transportation, such as buses, trains, and subways, is a popular alternative to driving, especially in urban areas. Public transportation produces fewer emissions per passenger than individual cars, which makes it a more environmentally-friendly option. However, public transportation also has its own environmental impacts, including the energy and resources required to build and maintain the infrastructure, and the emissions produced by the vehicles themselves.

Walking and biking

Walking and biking are the most environmentally-friendly modes of transportation, producing no emissions and requiring no energy beyond human power. However, walking and biking are not always practical, especially for longer distances or in areas without safe infrastructure for pedestrians and cyclists.

Comparing the environmental impact of electric vehicles to other modes of transportation is complex, as each mode has its own benefits and limitations. However, studies have shown that electric vehicles can be a significant improvement over gasoline-powered vehicles in terms of emissions, even when taking into account the emissions produced during the manufacturing of the vehicles and their batteries.

For example, a 2015 study by the Union of Concerned Scientists found that electric vehicles produce fewer emissions than gasoline-powered vehicles in all parts of the country, even in areas where the electricity grid is powered by coal-fired power plants. The study found that the average electric vehicle produces less than half the global warming emissions of the average gasoline-powered vehicle.

In addition, a 2017 study by the Norwegian University of Science and Technology found that electric vehicles have lower life cycle greenhouse gas emissions than gasoline-powered vehicles, even when taking into account the emissions produced during the manufacturing of the vehicles and their batteries. The study found that electric vehicles produce about 50% less greenhouse gas emissions than gasoline-powered vehicles on average.

Overall, while electric vehicles are not without their environmental impacts, they have the potential to be a significant improvement over gasoline-powered vehicles, especially when powered by renewable energy sources. However, to fully realize the environmental benefits of electric vehicles, it is important to also consider the impact of the manufacturing and disposal of the vehicles and their components, as well as the impact of the energy sources used to power them.

Chapter 5: Electric Vehicles and the Future of Transportation

The role of electric vehicles in reducing emissions

As the world continues to grapple with climate change and the need to reduce carbon emissions, the role of electric vehicles in the future of transportation cannot be overstated. Electric vehicles have the potential to significantly reduce greenhouse gas emissions, particularly if they are powered by renewable energy sources. In this chapter, we will explore the role of electric vehicles in reducing emissions and the potential benefits for the environment.

Electric vehicles and greenhouse gas emissions

One of the main reasons why electric vehicles are considered an important tool in reducing greenhouse gas emissions is because they do not emit tailpipe emissions like their traditional combustion engine counterparts. Instead of burning fossil fuels, electric vehicles rely on electricity from the grid or other renewable sources like solar or wind power. This means that they produce no emissions directly from the vehicle itself.

However, it is important to note that electric vehicles are not entirely emissions-free. The emissions associated with electric vehicles depend on the source of electricity used to charge them. In regions where the electricity comes from coal-fired power plants, electric vehicles may produce emissions equivalent to a traditional gasoline-powered car. However, as the electricity grid becomes cleaner, the emissions associated with electric vehicles will continue to decrease.

In addition to reducing tailpipe emissions, electric vehicles can also help to reduce overall emissions by improving the efficiency of the transportation system. Electric vehicles are more energy-efficient than traditional combustion engine vehicles and can be designed to recover energy through regenerative braking. They also have the potential to be used as energy storage devices, helping to balance the grid and store excess renewable energy during times of low demand.

The benefits of reducing emissions

Reducing greenhouse gas emissions has numerous benefits for the environment and human health. By reducing emissions, we can help to mitigate the impacts of climate change, such as rising sea levels, more severe weather events, and increased risk of wildfires. Reducing emissions can also improve air quality, which has been linked to a range of health issues, including asthma, lung cancer, and cardiovascular disease.

In addition to environmental and health benefits, reducing emissions can also have economic benefits. Electric vehicles can help to reduce our reliance on imported oil and increase energy security. They can also create new jobs in the clean energy sector and stimulate economic growth.

The role of electric vehicles in reducing emissions

Electric vehicles have the potential to play a significant role in reducing greenhouse gas emissions in the transportation sector. According to the International Energy Agency (IEA), electric vehicles could represent nearly one-third of the global passenger car fleet by 2030, and up to 60% of the fleet by 2040.

To achieve these levels of adoption, however, there are several barriers that need to be addressed. One of the main barriers is the cost of electric vehicles, which is still higher than traditional combustion engine vehicles. However, as battery technology continues to improve and production scales up, the cost of electric vehicles is expected to decrease.

Another barrier is the availability of charging infrastructure. Many people are hesitant to buy electric vehicles because they are concerned about running out of charge on long trips or being unable to find a charging station. To address this issue, governments and private companies are investing in the development of charging infrastructure, including fast-charging stations along highways and in urban areas.

Conclusion

Electric vehicles have the potential to play a significant role in reducing greenhouse gas emissions and improving the efficiency of the transportation system. While there are still barriers to widespread adoption, the continued improvement of battery technology and the expansion of charging infrastructure are making electric vehicles a more viable option for consumers. As we continue to grapple with climate change and the need to reduce emissions, the role of electric vehicles in the future of transportation will become increasingly important.

The growth of electric vehicle adoption

Introduction: Electric vehicles are becoming an increasingly popular choice among consumers, thanks to their environmental benefits, cost savings, and technological advancements. In this chapter, we will discuss the growth of electric vehicle adoption and the factors driving this trend.

Section 1: Global Electric Vehicle Sales The global electric vehicle market has experienced rapid growth in recent years. According to the International Energy Agency (IEA), sales of electric cars surpassed two million units in 2018, representing a 72% increase from 2017. China, the United States, and Europe are the largest markets for electric vehicles, accounting for more than 90% of global electric vehicle sales in 2018. The adoption of electric vehicles is expected to continue to grow, with the IEA projecting that there will be 125 million electric vehicles on the road by 2030.

Section 2: Government Policies and Incentives Government policies and incentives play a significant role in promoting the adoption of electric vehicles. Many countries have implemented policies such as tax incentives, subsidies, and rebates to encourage consumers to switch to electric vehicles. In addition, some countries have implemented stricter emissions regulations, which are driving automakers to develop more electric vehicle models.

Section 3: Technological Advancements Technological advancements have made electric vehicles more affordable, reliable, and convenient for consumers. Battery technology has improved significantly, enabling electric vehicles to travel longer distances on a single charge. In addition, the development of fast-charging infrastructure has made it easier for consumers to recharge their electric vehicles on the go. Electric vehicle manufacturers have also been introducing new models with longer ranges and faster charging times, which are more appealing to consumers.

Section 4: Consumer Awareness and Perception As consumers become more aware of the environmental and cost-saving benefits of electric vehicles, their perception of electric vehicles is changing. Electric vehicles are no longer seen as a niche product, but rather as a viable alternative to traditional gasoline-powered vehicles. The growing popularity of electric vehicles has led to more media coverage and consumer education, which has contributed to a positive perception of electric vehicles among the public.

Section 5: Challenges and Opportunities Despite the growth in electric vehicle adoption, there are still challenges that need to be addressed. One of the biggest challenges is the high cost of electric vehicles compared to gasoline-powered vehicles. In addition, the lack of charging infrastructure in some areas can be a barrier to electric vehicle adoption. However, these challenges also present opportunities for innovation and investment in the electric vehicle industry. As more consumers switch to electric vehicles, there will be opportunities for companies to develop new products and services, such as charging stations and battery recycling facilities.

Conclusion: The growth of electric vehicle adoption is a positive trend for the environment, the economy, and consumers. Government policies, technological advancements, consumer awareness, and perception are all driving factors behind this trend. While there are still challenges that need to be addressed, the opportunities for innovation and investment in the electric vehicle industry are vast. As more consumers switch to electric vehicles, the future of transportation looks brighter than ever before.

The impact of electric vehicles on the energy grid

Electric vehicles (EVs) have been touted as a solution to reduce greenhouse gas emissions and mitigate climate change. However, their increasing popularity has also raised concerns about their impact on the energy grid. As more EVs are being produced and sold, it is essential to examine the effects of this transition on the energy system.

The integration of EVs into the energy grid presents both opportunities and challenges. EVs can help reduce the overall demand for fossil fuels by using electricity generated from renewable sources, such as wind and solar power. Moreover, EVs can act as a distributed energy resource by providing energy storage capabilities through their batteries, which can be used to balance the grid during periods of high demand.

However, the rapid growth of EVs also presents challenges to the energy grid. One of the most significant challenges is the need for charging infrastructure. As the number of EVs on the road increases, the demand for charging stations will also increase. This, in turn, will place a strain on the energy grid, especially during peak demand periods. It will require significant investment in charging infrastructure and upgrading the energy grid to handle the increased demand.

Another challenge is the impact of EVs on the grid's stability. EVs are expected to increase the demand for electricity significantly, which can lead to voltage fluctuations and grid instability, especially in areas with a high concentration of EVs. This issue can be addressed by implementing smart charging systems, which can balance the demand for electricity with the grid's capacity.

Moreover, the growth of EVs could lead to an increase in energy consumption. While EVs are more energy-efficient than conventional vehicles, the increased demand for electricity could lead to an overall increase in energy consumption. This, in turn, could increase the need for additional power plants or the use of non-renewable sources of energy to meet the demand.

Another concern is the potential for EVs to strain the grid during extreme weather events. In areas where the grid is already under stress due to high temperatures or severe weather conditions, the increased demand for electricity from EVs could lead to blackouts or brownouts.

To address these challenges, policymakers and industry leaders need to develop strategies that promote the integration of EVs into the energy system while maintaining grid stability and reliability. These strategies could include:

1. Building a robust charging infrastructure network: Building a network of charging stations across the country will be essential to support the growing number of EVs on the road. Governments and industry leaders need to invest in building charging stations, especially in urban areas.

2. Implementing smart charging systems: Smart charging systems can balance the demand for electricity with the grid's capacity, reducing the strain on the energy system during peak demand periods. This will require the integration of advanced software systems and smart metering technology.

3. Increasing the use of renewable energy sources: The use of renewable energy sources, such as wind and solar power, can help reduce the overall demand for fossil fuels and promote a more sustainable energy system.

4. Encouraging the development of energy storage solutions: The development of energy storage solutions, such as large-scale batteries, can help address the issue of grid instability by providing energy storage capabilities during periods of high demand.

5. Promoting energy-efficient building design: Energy-efficient building design can reduce the overall demand for electricity, thereby reducing the strain on the energy grid.

In conclusion, the growth of EVs presents both opportunities and challenges to the energy grid. While EVs can help reduce greenhouse gas emissions and promote a more sustainable energy system, their rapid growth could strain the energy grid's capacity and stability. Governments and industry leaders need to develop strategies that promote the integration of EVs into the energy system while maintaining grid reliability and stability. This can include measures such as smart charging infrastructure, time-of-use pricing, vehicle-to-grid technologies, and investments in renewable energy sources. The integration of these solutions will require collaboration between various stakeholders, including utilities, automakers, policymakers, and consumers. Additionally, advancements in battery technology and energy storage solutions will also play a crucial role in ensuring the successful integration of EVs into the energy grid. With the right strategies and investments, the growth of EVs can be harnessed to build a cleaner, more resilient, and sustainable energy system for the future.

Chapter 6: Challenges Facing Electric Vehicles
Battery range and charging infrastructure

Battery range and charging infrastructure are two of the biggest challenges facing electric vehicles (EVs) today. While advancements in battery technology have significantly increased the range of EVs over the years, they still have limited ranges compared to conventional vehicles. Most electric cars can travel between 100 and 300 miles on a single charge, depending on the model and battery capacity. This range may be sufficient for some drivers' daily commutes or short trips, but it is often insufficient for long-distance travel or for those who cannot charge their cars at home.

To address this challenge, automakers are working to develop more advanced batteries with longer ranges. Solid-state batteries, for example, have the potential to double the range of EVs while reducing charging times. Solid-state batteries replace the liquid electrolyte used in conventional lithium-ion batteries with a solid material, which improves energy density and reduces the risk of fires. Toyota plans to introduce solid-state batteries in its electric cars by 2025.

Another solution is to increase the number of charging stations available to EV drivers. The availability of charging infrastructure is critical to the widespread adoption of EVs. While the number of charging stations has increased significantly in recent years, there is still a significant gap in the availability of charging infrastructure, particularly in rural areas. The development of fast-charging technology, which can recharge an EV in as little as 30 minutes, has also helped address this challenge. Fast-charging stations are becoming more common on major highways and in urban areas.

Another challenge related to charging infrastructure is the need for standardization. Different automakers use different charging systems, making it difficult for EV drivers to find compatible charging stations. The establishment of charging standards and protocols is essential to ensure that all EVs can use any charging station, regardless of the automaker or model.

Additionally, the affordability of EVs remains a challenge. While the cost of EVs has decreased significantly in recent years, they are still more expensive than traditional gasoline vehicles. The initial purchase price of an EV is often higher, and battery replacement costs can also be significant. However, the total cost of ownership of an EV is typically lower than that of a gasoline vehicle due to lower operating and maintenance costs.

Another challenge is the limited availability of models and styles. While the number of EV models available has increased in recent years, there are still fewer options than for gasoline vehicles. Many automakers are prioritizing the development of SUVs and trucks, which are more profitable, over smaller, more affordable EVs.

Finally, the lack of public awareness and education about EVs is also a challenge. Many people are unfamiliar with EVs and have misconceptions about their performance and capabilities. Governments and industry leaders need to work to increase public awareness and education about the benefits and potential of EVs to accelerate their adoption.

In conclusion, battery range and charging infrastructure remain two of the most significant challenges facing the widespread adoption of EVs. While advancements in battery technology and charging infrastructure have been made, there is still a need for further development and standardization to improve EVs' convenience and accessibility. Governments and industry leaders must work together to address these challenges and promote the adoption of EVs as a critical component of a more sustainable transportation system.

Consumer adoption and education are significant challenges facing electric vehicles (EVs). Despite the growth in the EV market in recent years, many consumers still have misconceptions and lack knowledge about EVs. In this chapter, we will explore the challenges associated with consumer adoption and education, as well as potential solutions.

One significant challenge with consumer adoption is the perceived higher cost of EVs compared to traditional gasoline-powered cars. While EVs may have a higher upfront cost, they have lower operational costs due to the lower cost of electricity compared to gasoline. It is crucial to educate consumers about the total cost of ownership, including factors such as fuel and maintenance costs, to make informed decisions when purchasing a vehicle.

Another challenge is range anxiety, which refers to the fear of running out of battery charge and being stranded without a charging station nearby. This concern is especially prevalent among consumers who live in areas with limited charging infrastructure. It is essential to educate consumers about the range capabilities of EVs and the growing network of charging stations.

Additionally, some consumers have concerns about the reliability and durability of EVs. Many people are familiar with gasoline-powered vehicles, which have been around for over a century, while EVs are relatively new technology. Education and awareness about the durability and reliability of EVs can help alleviate these concerns.

Moreover, consumers may have limited knowledge about the different types of EVs available, including battery electric vehicles (BEVs), plug-in hybrid electric vehicles (PHEVs), and hybrid electric vehicles (HEVs). It is essential to provide consumers with accurate and accessible information about the various EV options to make informed purchase decisions.

To address these challenges, various initiatives are underway to increase consumer education and adoption of EVs. Governments, non-profit organizations, and automakers are partnering to provide incentives, education campaigns, and infrastructure development. For instance, some cities have launched campaigns to educate consumers about the benefits of EVs and increase awareness about charging infrastructure. In addition, automakers are investing in research and development to improve the performance, affordability, and range of EVs.

In conclusion, consumer adoption and education are significant challenges facing the widespread adoption of EVs. These challenges can be addressed through education campaigns, infrastructure development, and incentives. As consumers become more informed about the benefits and capabilities of EVs, we can expect to see more widespread adoption and a shift towards a more sustainable transportation system.

Cost and accessibility

Electric vehicles (EVs) have the potential to transform the transportation sector by reducing greenhouse gas emissions and dependence on fossil fuels. However, despite their numerous benefits, there are still several challenges facing EVs. One of the most significant challenges is the cost and accessibility of EVs.

Cost of Electric Vehicles

Electric vehicles are often more expensive than conventional gasoline vehicles, primarily due to the cost of the battery. The battery is the most expensive component of an EV, and its cost can range from $5,000 to $20,000 or more depending on the size and range of the battery. The high cost of batteries makes EVs more expensive than gasoline vehicles, which can deter some consumers from purchasing them.

Additionally, EVs may require specialized maintenance and repairs, which can also be more expensive than those for gasoline vehicles. For example, if an EV battery needs to be replaced, it can cost several thousand dollars, while replacing a gasoline engine may cost less.

However, the cost of EVs is gradually decreasing due to advancements in battery technology and increased production volumes. In the future, it is expected that the cost of EVs will continue to decline, making them more affordable for consumers.

Accessibility of Electric Vehicles

Another challenge facing the widespread adoption of EVs is their accessibility. EV charging infrastructure is not yet as widespread as gasoline refueling infrastructure, which can make it difficult for consumers to find convenient and reliable places to charge their vehicles. This is especially true for people who live in apartments or do not have access to off-street parking, as they may not have a convenient place to charge their vehicle at home.

Additionally, some consumers may not have the financial means to purchase an EV, which can limit the accessibility of these vehicles to certain demographics. Although there are incentives and tax credits available to help offset the cost of EVs, these programs may not be accessible to all consumers.

The lack of accessibility of EVs can also lead to a "chicken and egg" problem. For example, if there are not enough EVs on the road, businesses may be less likely to invest in EV charging infrastructure, which in turn can limit the adoption of EVs.

Solutions to Cost and Accessibility Challenges

To address the cost and accessibility challenges facing EVs, there are several solutions that governments and industry leaders can consider.

One solution is to offer incentives and tax credits to consumers to make EVs more affordable. For example, some states in the United States offer rebates for the purchase of EVs, while some countries, such as Norway, offer significant tax incentives for EVs. These programs can help reduce the initial cost of purchasing an EV and make them more accessible to a wider range of consumers.

Another solution is to invest in EV charging infrastructure to make it more convenient and accessible for consumers. This can include installing more public charging stations in urban areas, as well as promoting the installation of charging stations in multi-unit dwellings and other areas where EV owners may not have access to a personal charging station.

Industry leaders can also work to reduce the cost of EVs through advancements in battery technology and increased production volumes. As the cost of batteries continues to decrease, the cost of EVs will also decrease, making them more accessible to a wider range of consumers.

Finally, education and awareness campaigns can help consumers understand the benefits of EVs and overcome any misconceptions they may have. By promoting the environmental and economic benefits of EVs, consumers may be more willing to consider purchasing an EV and investing in the necessary charging infrastructure.

Conclusion

The cost and accessibility of EVs are significant challenges facing the widespread adoption of these vehicles. However, there are solutions that can help reduce the cost of EVs and make them more accessible to a broader range of consumers. One solution is government incentives, such as tax credits and rebates, for purchasing EVs. Governments can also invest in the development of charging infrastructure to make EVs more accessible and convenient for consumers. Additionally, automakers can focus on increasing production and efficiency to reduce the cost of EVs over time. The rise of shared mobility, such as ride-sharing and car-sharing services, can also make EVs more accessible and affordable for consumers who may not be able to afford their own vehicle. Ultimately, addressing the cost and accessibility challenges of EVs requires collaboration between governments, automakers, and other stakeholders to create a more sustainable and equitable transportation system.

Emerging battery technologies

Electric vehicles (EVs) are rapidly gaining popularity due to their potential to revolutionize the transportation sector by providing clean, sustainable, and efficient transportation. One of the key components of an electric vehicle is its battery, which stores energy for the electric motor to use. The battery's performance and capabilities can have a significant impact on the vehicle's range, charging time, and overall efficiency. Therefore, the development of new and improved battery technologies is crucial for the continued growth and success of electric vehicles.

In recent years, there have been several emerging battery technologies that show great potential to overcome some of the limitations of current lithium-ion batteries used in most electric vehicles. These technologies include solid-state batteries, lithium-sulfur batteries, and flow batteries.

Solid-state batteries are considered one of the most promising emerging technologies for electric vehicles. Unlike current lithium-ion batteries that use liquid electrolytes, solid-state batteries use solid electrolytes, which can improve energy density and safety while reducing the risk of leakage or fire. Solid-state batteries can also potentially provide faster charging times, longer lifespan, and improved performance in extreme temperatures.

Lithium-sulfur batteries are another promising technology that could revolutionize the electric vehicle industry. They offer a higher theoretical energy density than lithium-ion batteries, which means they can potentially provide longer ranges. Lithium-sulfur batteries are also cheaper and more environmentally friendly to produce, as sulfur is abundant and non-toxic compared to the cobalt and nickel used in lithium-ion batteries. However, lithium-sulfur batteries face challenges such as lower cycle life and higher self-discharge rates, which researchers are currently working to overcome.

Flow batteries are a type of rechargeable battery that uses two separate liquids to store energy. Flow batteries can potentially provide high energy density, long cycle life, and the ability to charge and discharge simultaneously, making them well-suited for electric vehicles. Flow batteries also have the added benefit of being easily rechargeable by simply refilling the liquid electrolytes, reducing the need for expensive and time-consuming battery replacements.

In addition to these emerging battery technologies, researchers are also exploring the potential of other materials, such as sodium-ion and magnesium-ion, to be used in electric vehicle batteries. Sodium and magnesium are abundant and inexpensive materials that could provide a more sustainable and cost-effective alternative to lithium-ion batteries.

However, despite the potential of these emerging battery technologies, there are still significant challenges that need to be addressed before they can be commercially viable. These challenges include improving the energy density, cycle life, and safety of these batteries while reducing their cost and making them scalable for mass production.

Another promising innovation in the field of electric vehicles is wireless charging. Wireless charging technology allows electric vehicles to charge without physically plugging them into a charging station. Instead, the vehicle's battery is charged using electromagnetic induction, which transfers energy wirelessly from a charging pad on the ground to a receiver coil on the vehicle's undercarriage. This technology has the potential to make charging more convenient and seamless for electric vehicle owners, as they would not need to physically plug and unplug their vehicles every time they need to charge.

Finally, the development of autonomous electric vehicles is another significant innovation that could revolutionize the transportation sector. Autonomous vehicles rely on advanced sensors and software to drive themselves, eliminating the need for human drivers. Electric autonomous vehicles have the potential to significantly reduce greenhouse gas emissions, traffic congestion, and the number of road accidents while providing more efficient and convenient transportation for passengers.

In conclusion, emerging battery technologies such as solid-state, lithium-sulfur, and flow batteries, as well as wireless charging and autonomous electric vehicles, show great promise in overcoming the challenges facing electric vehicles. Continued research and development of these technologies will be crucial for the widespread adoption of electric vehicles and the transition to a more sustainable and efficient transportation system.

Autonomous driving technology

Autonomous driving technology is an emerging innovation in the field of electric vehicles that has the potential to revolutionize the way we commute and travel. Autonomous driving refers to the ability of a vehicle to operate without the need for human intervention. This technology is being developed by several automakers, including Tesla, Google, and Uber, and is expected to be widely available in the coming years.

One of the main benefits of autonomous driving technology is improved safety on the roads. Self-driving cars are equipped with advanced sensors, cameras, and machine learning algorithms that enable them to detect and respond to potential hazards on the road. This can significantly reduce the number of accidents caused by human error, which is currently the leading cause of traffic fatalities.

Another benefit of autonomous driving technology is increased efficiency and reduced congestion on the roads. Self-driving cars can communicate with each other and with traffic signals to optimize their routes and avoid congestion. This can lead to shorter travel times, less time spent stuck in traffic, and a more efficient use of road space.

In addition to these benefits, autonomous driving technology can also improve accessibility and mobility for people with disabilities or those who are unable to drive. Self-driving cars can provide a safe and convenient mode of transportation for these individuals, allowing them to travel independently and without assistance.

Despite the potential benefits of autonomous driving technology, there are also several challenges that need to be addressed. One of the main challenges is the need for robust and reliable software that can ensure the safe operation of self-driving cars. This requires significant investment in research and development, as well as testing and validation to ensure the software is reliable and free from bugs or vulnerabilities.

Another challenge is the need for a regulatory framework that can ensure the safe and responsible deployment of autonomous driving technology. Governments and regulatory bodies need to establish standards and regulations that ensure self-driving cars meet safety and performance requirements, and that they are operated in a responsible manner.

In conclusion, autonomous driving technology is an exciting innovation in the field of electric vehicles that has the potential to revolutionize the way we commute and travel. While there are several challenges that need to be addressed, the benefits of self-driving cars are clear, including improved safety, increased efficiency, and improved accessibility and mobility. As this technology continues to evolve, it will be important for governments, industry leaders, and consumers to work together to ensure its safe and responsible deployment.

Integrating electric vehicles into smart cities

As cities around the world work to become more sustainable and reduce their carbon footprints, the role of electric vehicles (EVs) is becoming increasingly important. One key trend in this area is the development of "smart cities," which integrate advanced technologies to improve efficiency, reduce waste, and enhance quality of life for residents. In this context, EVs can play a major role as a cleaner and more efficient form of transportation. In this section, we'll explore some of the ways that EVs are being integrated into smart cities, and the benefits this integration can bring.

What are smart cities?

Before we dive into the specifics of EV integration, let's define what we mean by "smart cities." The term refers to cities that use advanced technologies to improve various aspects of urban life, such as:

- Energy efficiency: Smart cities often rely on renewable energy sources like solar and wind power, and use technologies like smart grids to optimize energy usage and reduce waste.

- Transportation: Smart cities may use advanced traffic management systems, public transportation networks, and even autonomous vehicles to reduce congestion and improve mobility.

- Public services: Advanced technologies can help cities deliver public services more efficiently, such as through smart waste management systems that reduce landfill waste and recycling programs that use sensors to optimize collection.

- Quality of life: Smart cities may offer amenities like free Wi-Fi, public green spaces, and bike-share programs to improve quality of life for residents.

The role of EVs in smart cities

So how do electric vehicles fit into this picture? There are a few key ways:

- Reduced emissions: EVs produce fewer emissions than gasoline-powered cars, which can help smart cities reduce their overall carbon footprint and improve air quality.

- Increased efficiency: EVs are more energy-efficient than gasoline cars, which means they can help smart cities optimize their energy usage and reduce waste.

- Improved mobility: By integrating EV charging infrastructure with public transportation networks, smart cities can provide residents with a seamless and convenient way to get around.

- Enhanced connectivity: EVs can be connected to smart grids, which can help cities manage energy usage more efficiently and reduce strain on the grid during peak hours.

EV charging infrastructure in smart cities

One of the key challenges to EV adoption is the availability of charging infrastructure. In smart cities, however, EV charging infrastructure can be integrated into the existing transportation network, making it more convenient and accessible for residents. For example:

- Public charging stations: Cities can install public charging stations at key locations like parking garages, shopping centers, and public transit hubs to provide convenient charging options for residents and visitors.

- Integrated charging networks: By integrating EV charging infrastructure with public transportation networks, smart cities can provide residents with a seamless and convenient way to get around. For example, a bus stop could have an EV charging station that can be used by both buses and personal vehicles.

- Dynamic pricing: Smart cities can use dynamic pricing to encourage EV drivers to charge their cars during off-peak hours, when demand for energy is lower. This can help reduce strain on the energy grid during peak hours and lower overall electricity costs.

Intelligent transportation systems (ITS)

Intelligent transportation systems (ITS) are another key component of smart cities. These systems use advanced technologies like sensors, cameras, and data analytics to optimize traffic flow and reduce congestion. When integrated with EVs, ITS can help:

- Improve EV battery range: By optimizing traffic flow, ITS can help EVs maintain a more consistent speed, which can improve battery range.

- Provide real-time traffic information: Smart cities can use ITS to provide real-time traffic information to EV drivers, helping them find the fastest and most efficient routes.

- Optimize EV charging: By analyzing traffic patterns, ITS can help smart cities optimize the placement and pricing of EV charging stations to reduce

Conclusion

The future of electric vehicles and sustainable transportation

The future of electric vehicles (EVs) and sustainable transportation is promising. With increasing concerns about climate change, air pollution, and dependence on fossil fuels, the transportation sector is shifting towards a more sustainable future. EVs are at the forefront of this shift, and their adoption is expected to continue to grow rapidly in the coming years. In this section, we will explore the future of EVs and sustainable transportation, including the challenges and opportunities that lie ahead.

One of the main opportunities for EVs is their potential to significantly reduce greenhouse gas emissions and air pollution. As more renewable energy sources are added to the grid, the environmental benefits of EVs will only increase. In addition, advances in battery technology and charging infrastructure will help alleviate the range anxiety that has been a significant barrier to EV adoption. As battery costs continue to decline, EVs will become more affordable, making them accessible to a broader range of consumers.

Another significant opportunity for EVs is their potential to integrate with emerging technologies, such as autonomous driving and the Internet of Things (IoT). These technologies could revolutionize the transportation sector and make it more efficient and sustainable. For example, autonomous driving could reduce traffic congestion and make EVs more practical for long-distance travel. The IoT could enable smart charging infrastructure, where EVs are charged during periods of low demand on the grid, reducing stress on the electrical system and lowering energy costs for consumers.

Despite the opportunities, there are also challenges that must be addressed to ensure the success of EVs and sustainable transportation. One of the most significant challenges is the need for a reliable and resilient energy grid. As more EVs are added to the grid, there is a risk of overloading the system during peak demand periods. This challenge can be addressed through the development of smart charging infrastructure and energy management systems that can balance energy demand and supply.

Another challenge is the need for robust policies and regulations that support EV adoption and sustainable transportation. Governments play a critical role in promoting the adoption of EVs through incentives, subsidies, and regulations that promote the development of charging infrastructure and the use of renewable energy sources. In addition, public education and awareness campaigns can help consumers understand the benefits of EVs and reduce the anxiety associated with the transition to electric transportation.

Finally, there is a need for continued innovation and research to drive the development of new technologies that can address the challenges of EV adoption and sustainable transportation. Emerging technologies, such as solid-state batteries and fuel cells, have the potential to overcome the limitations of current battery technology and enable longer-range EVs with shorter charging times. In addition, advances in materials science and engineering could lead to the development of lighter and more efficient EV components, further reducing the cost and improving the performance of EVs.

In conclusion, the future of EVs and sustainable transportation is bright. EVs have the potential to significantly reduce greenhouse gas emissions and air pollution, while also improving energy efficiency and reducing dependence on fossil fuels. However, achieving this vision requires addressing the challenges associated with EV adoption, including the need for a reliable energy grid, robust policies and regulations, and continued innovation and research. Through collaboration between industry, government, and consumers, we can work towards a more sustainable transportation system and a cleaner future for all.

The role of individuals and society in promoting sustainable transportation

Introduction: The adoption of sustainable transportation is crucial for reducing greenhouse gas emissions and mitigating the impacts of climate change. Electric vehicles (EVs) are one of the key components of sustainable transportation, and their widespread adoption is essential for achieving a more sustainable transportation system. However, the role of individuals and society in promoting sustainable transportation is equally important. In this section, we will discuss the ways in which individuals and society can play a critical role in promoting sustainable transportation.

Individuals and Sustainable Transportation: Individuals play a critical role in promoting sustainable transportation. The most obvious way that individuals can promote sustainable transportation is by choosing to drive an EV or use other sustainable transportation modes such as biking, walking, or public transportation. However, there are other ways that individuals can promote sustainable transportation as well. For example, individuals can advocate for policies that support sustainable transportation such as increased funding for public transportation, bike lanes, and EV charging infrastructure. They can also encourage their friends, family, and colleagues to adopt sustainable transportation habits and help raise awareness about the benefits of sustainable transportation.

Society and Sustainable Transportation: Society as a whole can also play a critical role in promoting sustainable transportation. One of the most important ways that society can promote sustainable transportation is by investing in infrastructure that supports sustainable transportation modes. This includes building bike lanes, expanding public transportation, and installing EV charging infrastructure. Society can also encourage the adoption of sustainable transportation by implementing policies that incentivize sustainable transportation, such as tax credits for EVs or subsidies for public transportation.

In addition to investing in infrastructure and implementing policies that support sustainable transportation, society can also promote sustainable transportation through education and awareness-raising efforts. By raising awareness about the benefits of sustainable transportation and the negative impacts of traditional transportation modes, society can help shift attitudes and behaviors towards sustainable transportation.

The Future of Sustainable Transportation: The future of sustainable transportation is bright, but it will require the collective efforts of individuals and society to make it a reality. The widespread adoption of EVs and other sustainable transportation modes is essential for reducing greenhouse gas emissions and mitigating the impacts of climate change. However, it will require investments in infrastructure, policies that support sustainable transportation, and education and awareness-raising efforts to make this a reality.

Conclusion: In conclusion, the role of individuals and society in promoting sustainable transportation is critical for achieving a more sustainable transportation system. Individuals can promote sustainable transportation by adopting sustainable transportation modes and advocating for policies that support sustainable transportation. Society can promote sustainable transportation by investing in infrastructure, implementing policies that incentivize sustainable transportation, and raising awareness about the benefits of sustainable transportation. By working together, we can create a more sustainable transportation system and help mitigate the impacts of climate change.

The importance of continued research and development in electric vehicles

Electric vehicles are a promising solution to reducing greenhouse gas emissions and mitigating the impacts of climate change. However, to fully realize the potential of electric vehicles, continued research and development are necessary. In this section, we will discuss the importance of continued research and development in electric vehicles and the key areas that require attention.

One of the key areas that require continued research and development is battery technology. Although lithium-ion batteries are currently the most widely used battery technology in electric vehicles, there are limitations to their performance, cost, and safety. Thus, researchers are exploring alternative battery chemistries, such as solid-state batteries, lithium-sulfur batteries, and sodium-ion batteries, which have the potential to offer higher energy density, longer life, and lower cost. Furthermore, battery recycling and reuse are also areas that require attention to address the environmental impact of batteries.

Another area that requires continued research and development is charging infrastructure. Although the number of charging stations has increased significantly in recent years, the availability, speed, and reliability of charging infrastructure are still significant barriers to the widespread adoption of electric vehicles. Thus, research is needed to develop new charging technologies, such as wireless charging and fast-charging, and to improve the integration of charging infrastructure with the electricity grid.

In addition to battery technology and charging infrastructure, there are several other areas that require attention to support the growth of electric vehicles. These include:

1. Vehicle-to-grid (V2G) technology: V2G technology allows electric vehicles to store and discharge energy back to the grid, which can help balance the grid and reduce the need for additional power plants. Research in this area is necessary to develop the technology and infrastructure to support V2G.

2. Lightweight materials: The weight of electric vehicles impacts their performance and range. Thus, researchers are exploring lightweight materials, such as carbon fiber and aluminum, to reduce the weight of electric vehicles and improve their efficiency.

3. Autonomous driving technology: Autonomous driving technology has the potential to improve the efficiency and safety of electric vehicles. However, research is needed

to develop the technology and regulations to support the widespread adoption of autonomous vehicles.

4. Integration with renewable energy sources: Electric vehicles can help support the integration of renewable energy sources, such as solar and wind, by storing excess energy and discharging it back to the grid when needed. Thus, research is needed to develop the technology and policies to support the integration of electric vehicles with renewable energy sources.

5. Consumer behavior: Consumer behavior plays a significant role in the adoption and use of electric vehicles. Thus, research is needed to understand consumer preferences and behavior, and to develop strategies to encourage the adoption and use of electric vehicles.

In conclusion, continued research and development are necessary to fully realize the potential of electric vehicles and support the transition to a more sustainable transportation system. Researchers, industry leaders, and policymakers must work together to address the key challenges facing electric vehicles and develop new technologies, policies, and strategies to support their growth. By investing in research and development, we can accelerate the transition to a cleaner, more efficient, and more sustainable transportation system.

THE END

Key Terms and Definitions

To help you better understand the language and concepts related to aging and older adults, below you will find a list of key terms and their definitions.

1. Electric Vehicle (EV): A vehicle that runs on electricity stored in its rechargeable batteries and electric motor.

2. Battery Electric Vehicle (BEV): An electric vehicle that solely runs on a battery-powered electric motor without any internal combustion engine.

3. Hybrid Electric Vehicle (HEV): A vehicle that uses both an electric motor and a gasoline or diesel engine to power its wheels.

4. Plug-in Hybrid Electric Vehicle (PHEV): A type of hybrid electric vehicle that can be charged through an external power source.

5. Range Anxiety: The fear or worry of an electric vehicle driver that the battery will run out of charge before reaching the destination.

6. Regenerative Braking: A system that uses the kinetic energy produced by braking to recharge the electric vehicle's battery.

7. Charging Station: A place where electric vehicles can be charged with electricity from the grid.

8. Electric Vehicle Supply Equipment (EVSE): The equipment used to supply electricity to electric vehicles for charging.

9. Lithium-ion battery: A type of rechargeable battery commonly used in electric vehicles.

10. Level 1, Level 2, Level 3 Charging: The three standard levels of charging for electric vehicles, with level 1 being the slowest and level 3 being the fastest.

Supporting Materials

Introduction

- International Energy Agency. (2020). Global EV Outlook
2020: Entering the decade of electric drive?
- United Nations Environment Programme. (2019).
Emissions Gap Report 2019.

Chapter 1

- Kirsch, D. (2000). The Electric Vehicle and the Burden of
History. Rutgers University Press.
- Wietschel, M., & Dütschke, E. (2016). Handbook of
Research on Mobility and Computing: Evolving Technologies
and Ubiquitous Impacts. IGI Global.

Chapter 2

- Gao, Y., Li, X., & Zhang, S. (2017). Handbook of Electric
Vehicles. Springer.
- National Renewable Energy Laboratory. (2020). Electric
Vehicle Basics.

Chapter 3

- Golubkov, A., & Makovik, I. (2019). Electric Powertrain:
Energy Systems, Power Electronics and Drives for Hybrid,
Electric and Fuel Cell Vehicles. Wiley.
- Pistoia, G. (2018). Electric and Hybrid Vehicles: Power
Sources, Models, Sustainability, Infrastructure and the
Market. Elsevier.

Chapter 4

- Chen, H., & Sun, F. (2019). Electric Vehicles and Energy Storage Systems: A Sustainable Future. CRC Press.
- National Academies of Sciences, Engineering, and Medicine. (2018). Reducing Fuel Consumption and Greenhouse Gas Emissions of Medium- and Heavy-Duty Vehicles, Phase Two: First Report.

Chapter 5

- International Energy Agency. (2021). Global EV Outlook 2021: Accelerating the Transition to Electric Vehicles.
- Lutsey, N., & Sperling, D. (2019). Three Revolutions in Urban Transportation. Island Press.

Chapter 6

- Kisacikoglu, M. C. (2020). Electric Vehicles and Smart Grids. Springer.
- National Renewable Energy Laboratory. (2020). Charging Electric Vehicles.

Chapter 7:

- Bifulco, G. N., Cicala, G., & D'Ambrosio, S. (2020). Electric Vehicle Technology Explained. Wiley.
- Kharazi, A., & Harirforoush, H. (2019). Intelligent Transportation and Planning: Breakthroughs in Research and Practice. IGI Global.

Conclusion:

- International Energy Agency. (2020). Energy Technology Perspectives 2020.

- Sperling, D., & Gordon, D. (2009). Two billion cars: driving toward sustainability. Oxford University Press.

www.ingramcontent.com/pod-product-compliance
Lightning Source LLC
LaVergne TN
LVHW011034200726
843509LV00011B/1277